ISAAC ASIMOV'S NEW LIBRARY OF THE UNIVERSE

GLOBAL SPACE PROGRAMS

BY ISAAC ASIMOV

WITH REVISIONS AND UPDATING BY FRANCIS REDDY

Gareth Stevens Publishing
MILWAUKEE

For a free color catalog describing Gareth Stevens' list of high-quality books, call 1-800-542-2595 (USA) or 1-800-461-9120 (Canada). Gareth Stevens' Fax: (414) 225-0377.

Library of Congress Cataloging-in-Publication Data

Asimov, Isaac.
Global space programs / by Isaac Asimov; with revisions and updating by Francis Reddy.
p. cm. — (Isaac Asimov's New library of the universe)
Rev. ed. of: The world's space programs. 1990.
Includes index.
Summary: Examines the contributions of various nations and cultures to the exploration of space.
ISBN 0-8368-1235-2
1. Space sciences—Juvenile literature. 2. Astronautics—Juvenile literature. [1. Space sciences. 2. Astronautics. 3. Outer space—Exploration.] I. Reddy, Francis, 1959-. II. Asimov, Isaac. World's space programs. III. Title. IV. Series: Asimov, Isaac. New library of the universe.
QB500.22.A82 1996
919.904—dc20 95-40366

This edition first published in 1996 by
Gareth Stevens Publishing
1555 North RiverCenter Drive, Suite 201
Milwaukee, Wisconsin 53212, USA

Project editor: Barbara J. Behm
Design adaptation: Helene Feider
Editorial assistant: Diane Laska
Production director: Teresa Mahsem
Picture research: Matthew Groshek and Diane Laska

Printed in the United States of America

1 2 3 4 5 6 7 8 9 99 98 97 96

To bring this classic of young people's information up to date, the editors at Gareth Stevens Publishing have selected two noted science authors, Greg Walz-Chojnacki and Francis Reddy. Walz-Chojnacki and Reddy coauthored the recent book *Celestial Delights: The Best Astronomical Events Through 2001.*

Walz-Chojnacki is also the author of the book *Comet: The Story Behind Halley's Comet* and various articles about the space program. He was an editor of *Odyssey*, an astronomy and space technology magazine for young people, for eleven years.

Reddy is the author of nine books, including *Halley's Comet*, *Children's Atlas of the Universe*, *Children's Atlas of Earth Through Time*, and *Children's Atlas of Native Americans*, plus numerous articles. He was an editor of *Astronomy* magazine for several years.

CONTENTS

We live in an enormously large place – the Universe. It's just in the last fifty-five years or so that we've found out how large it probably is. It's only natural that we would want to understand the place in which we live, so scientists have developed instruments – such as radio telescopes, satellites, probes, and many more – that have told us far more about the Universe than could possibly be imagined.

We have seen planets up close. We have learned about quasars and pulsars, black holes, and supernovas. We have gathered amazing data about how the Universe may have come into being and how it may end. Nothing could be more astonishing.

Human beings are at the beginning of an era of working and living in space. The United States, Russia, European nations, Japan, and other countries are involved in programs that will unite our planet in the vast endeavor of space exploration.

Isaac Asimov

Space Age Rocketry

Rockets have existed in the form of fireworks for centuries. But Space Age rocketry began in 1903. That is when Russian schoolteacher Konstantin Tsiolkovsky developed the mathematics of rocket flight.

An American scientist, Robert H. Goddard, fired the first liquid-fueled rocket in 1926. In 1942, a German war scientist, Wernher von Braun, developed the first large rocket, the V-2. V-2s could travel for hundreds of miles before reaching their target. With this, Space Age rockets had arrived.

After World War II, the United States and the former Soviet Union began a race to see which would be the first to conquer space. The Soviets launched the first satellite, *Sputnik*, in 1957. In 1961, Soviet cosmonaut Yuri Gagarin became the first man in space. Two years later, Soviet cosmonaut Valentina Tereshkova became the first woman in space. Over time, both the United States and the former Soviet Union sent probes to Venus, and U.S. probes also visited Mercury and Mars. In 1969, U.S. astronaut Neil Armstrong became the first person to walk on the Moon.

Opposite: The mighty *Saturn V* rocket (U.S.), one of the most powerful ever built, blasts off.

Below: In a liquid-fueled rocket, oxygen and fuel from storage tanks flow into the combustion chamber *(1)*, burn, and create hot gases that race out of the engine through a nozzle *(2)*. A guidance system *(3)* keeps the rocket on path. The hot gases escaping from the nozzle push on the rocket, sending it forward . . . just as the escaping gas of a balloon sends it forward *(4)*.

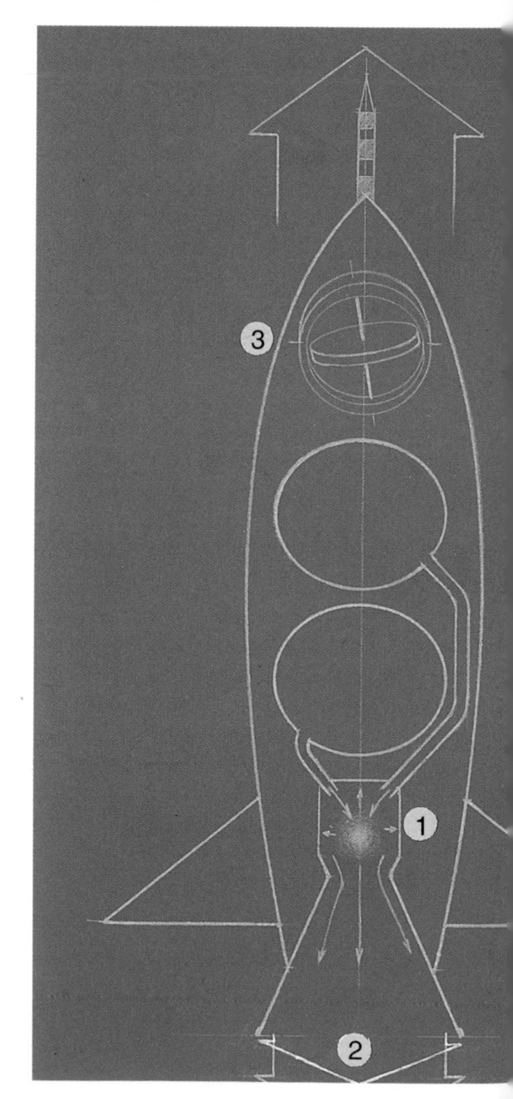

4
USA

Journey to the Moon

The U.S. Apollo program sent astronauts to the Moon six times between 1969 and 1972. Each mission put two astronauts on the Moon.

About 850 pounds (386 kilograms) of Moon rocks were brought back to Earth to be studied for clues about the Moon's early history. Since the Moon is a world without wind, water, or erosion, the rocks on its surface have not changed for billions of years. From them, scientists are discovering more about what Earth was like ages ago.

The last Apollo Moon flight was in 1972. Since then, no one has set foot on the Moon. Does that mean the world has abandoned exploration of the Moon? Definitely not!

Opposite: U.S. astronaut Edwin Aldrin, Jr., on the surface of the Moon.

Below: A flight to the Moon – a rocket blasts off *(1)*. The crew portion of the spacecraft joins its lunar lander craft after blasting out of Earth's orbit toward the Moon *(2)*. After entering Moon orbit, the vehicles separate. The lander drops to the surface *(3)*, while the other continues in orbit *(4)*. After the lunar crew has explored the Moon, the lander blasts off *(5)*, and the two crafts again join *(6-out of view)*. The lunar lander is then discarded. The crew leaves lunar orbit *(7)*. The crew capsule returns to Earth *(8)*.

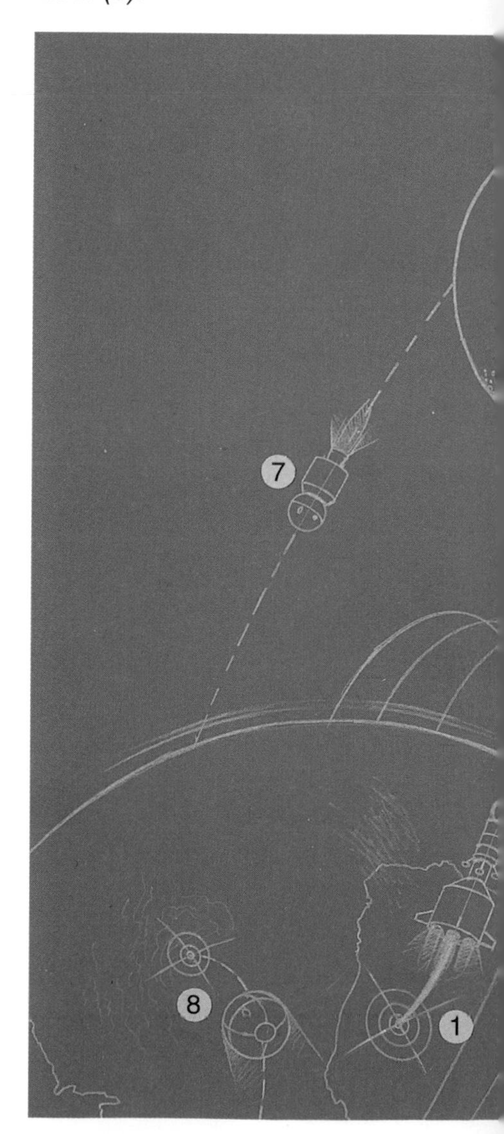

! *The rocket man*

Rocket pioneer Robert H. Goddard spent years designing, building, and testing rockets. He even drew up plans for sending a rocket to the Moon. During World War II, he tried to persuade the United States government to use his rockets as missiles. When the war ended, U.S. officials studying the German V-2 rocket asked German scientists how they designed it. "We read all of Goddard's books," they said. But it was too late for Goddard to know – he died just five days before the war ended.

6
5
4
3
2

A Parade of Satellites

Thousands of artificial satellites have been launched into orbit by the United States, Russia, the European Space Agency, Canada, China, India, and other nations. Today, artificial satellites come in all shapes and sizes. And they have many jobs to do.

Some study Earth itself – the health of forests and farms, or the location of fish and oil deposits. Weather satellites study the atmosphere, photographing Earth's clouds and tracking dangerous storms. Navigational satellites guide ships and aircraft. Spy satellites watch for military movement with cameras so sensitive they can read a license plate from 100 miles (160 kilometers) above Earth. Space telescopes study the Universe from positions in the cosmos for a crystal clear view.

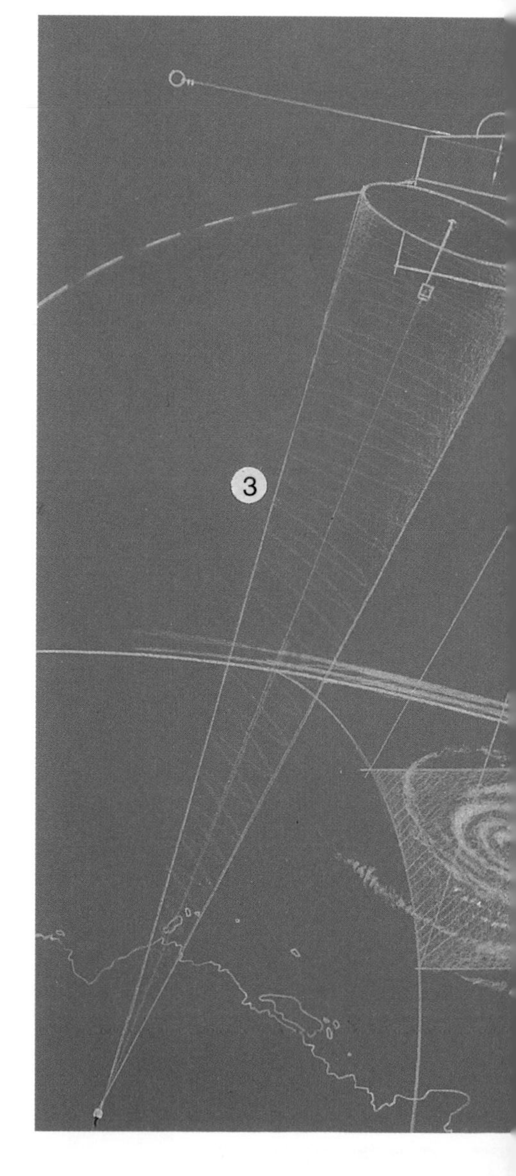

Opposite: Europe's *Meteosat* weather satellite monitors weather patterns from 23,000 miles (37,000 km) up.

Inset: The swirling clouds of a hurricane *(1)* are spotted by an orbiting weather satellite *(2)*. The satellite tracks and studies the storm and beams pictures and data to scientists on Earth *(3)*. This information gives scientists time to warn people threatened by such storms.

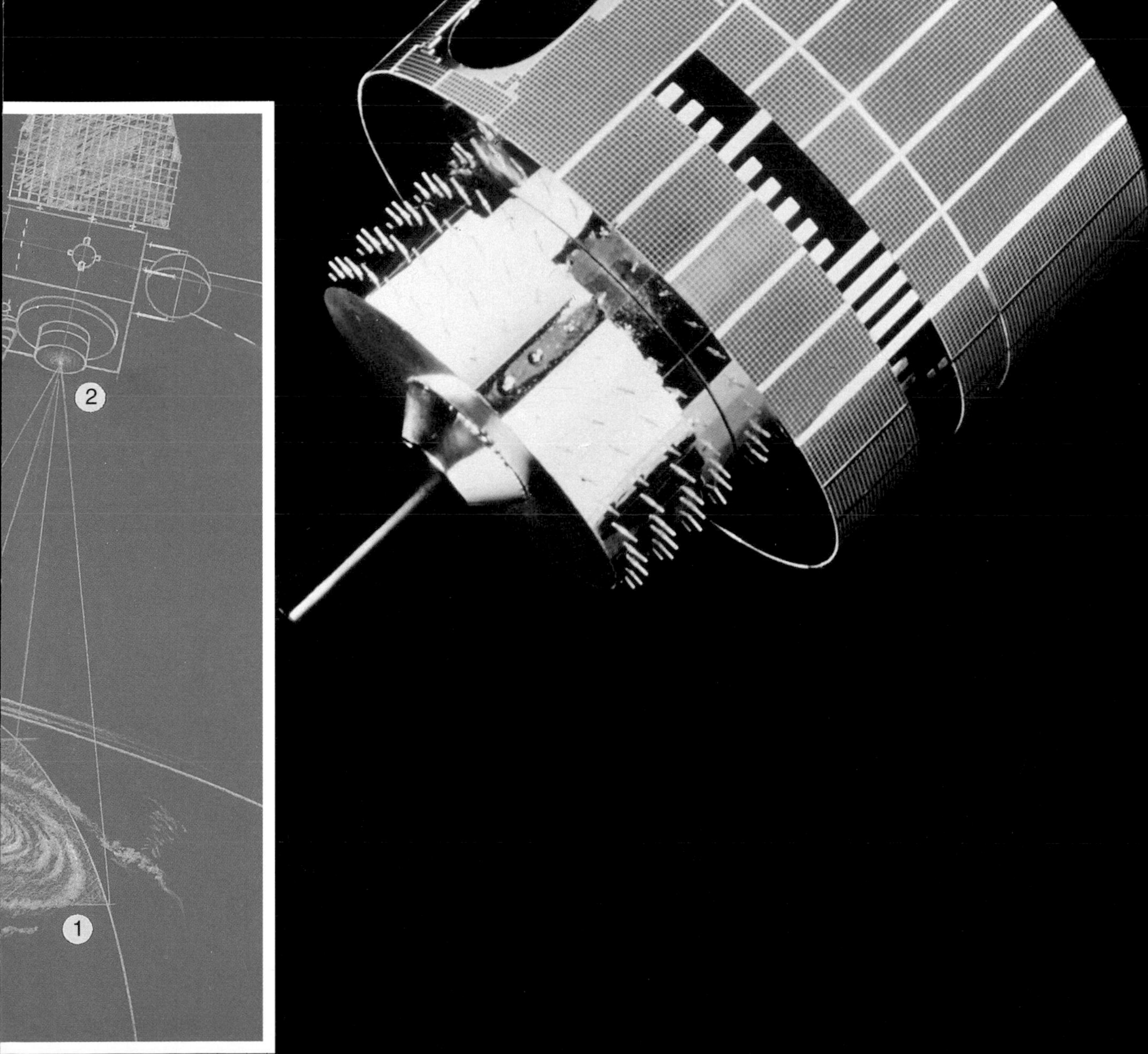
2
1

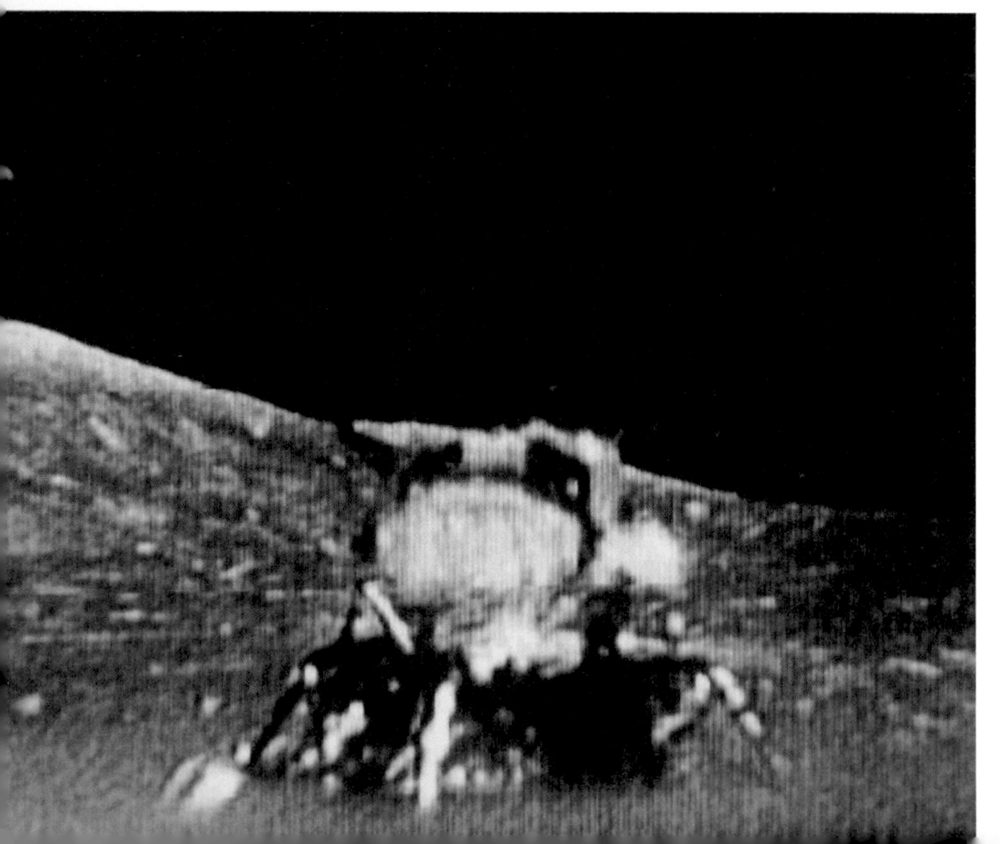

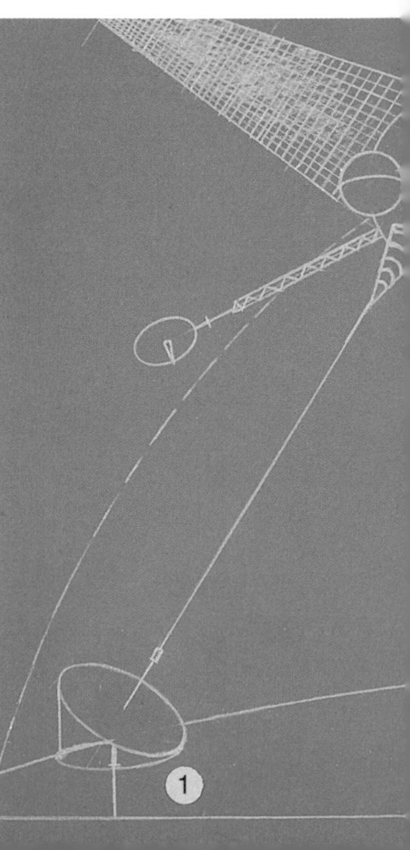
1

Modern Communications

Without artificial satellites, our modern-day communications would be impossible. Because of satellites, today's radio, telephone, and television systems have a global reach. The Cable News Network (CNN), for example, can be seen in over two hundred countries.

Communications satellites relay signals across oceans and continents. This allows millions of people to see and hear events that are happening in every corner of the world.

Canada and the nations of western Europe worked together to launch a communications satellite in 1989. Countries that do not have the ability to launch their own communications satellites can buy launch services from other countries.

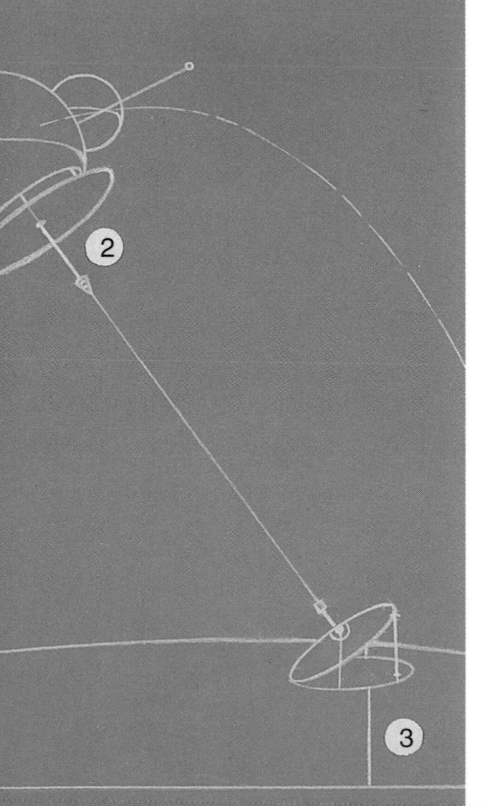

Top: A ground station linked to a communications satellite.

Opposite, bottom, left: Apollo 17 astronauts blast-off from the Moon. But if they're in the spacecraft, who's taking this picture? The camera was controlled from Earth by radio signals. A transmitter on the Moon returned the signals in the same way.

Left: A ground station on Earth *(1)* transmits TV or radio programs to an orbiting satellite *(2)*. The satellite receives the signals and beams them to another ground station *(3)* thousands of miles from the first.

Space Shuttles

The equipment used for launching humans into space is expensive. Because of this, space engineers worked for years to develop reusable spacecraft called space shuttles.

The first shuttle was launched in 1981 by the United States. Despite minor problems, the shuttle program was a great success. The European Space Agency and Russia have also developed shuttles, but their shuttle programs have not yet been successful.

The U.S. shuttles were crucial in repairing damaged satellites in orbit, and the shuttles launched new satellites from space. In addition, astronauts have carried out numerous experiments on board shuttles in the zero gravity of space. It was on the space shuttles that the first American women astronauts, Sally Ride and Judith Resnik, reached space.

Then, in January 1986, the space shuttle *Challenger* exploded during takeoff, killing all seven people aboard. NASA spent the next two years improving the shuttle and making it safer. In 1988, U.S. shuttle flights began again.

Opposite: In December 1993, shuttle astronauts took on their most challenging space work ever – fixing the Hubble Space Telescope. They attached new solar panels and repaired a flaw in the telescope's main mirror.

NASA

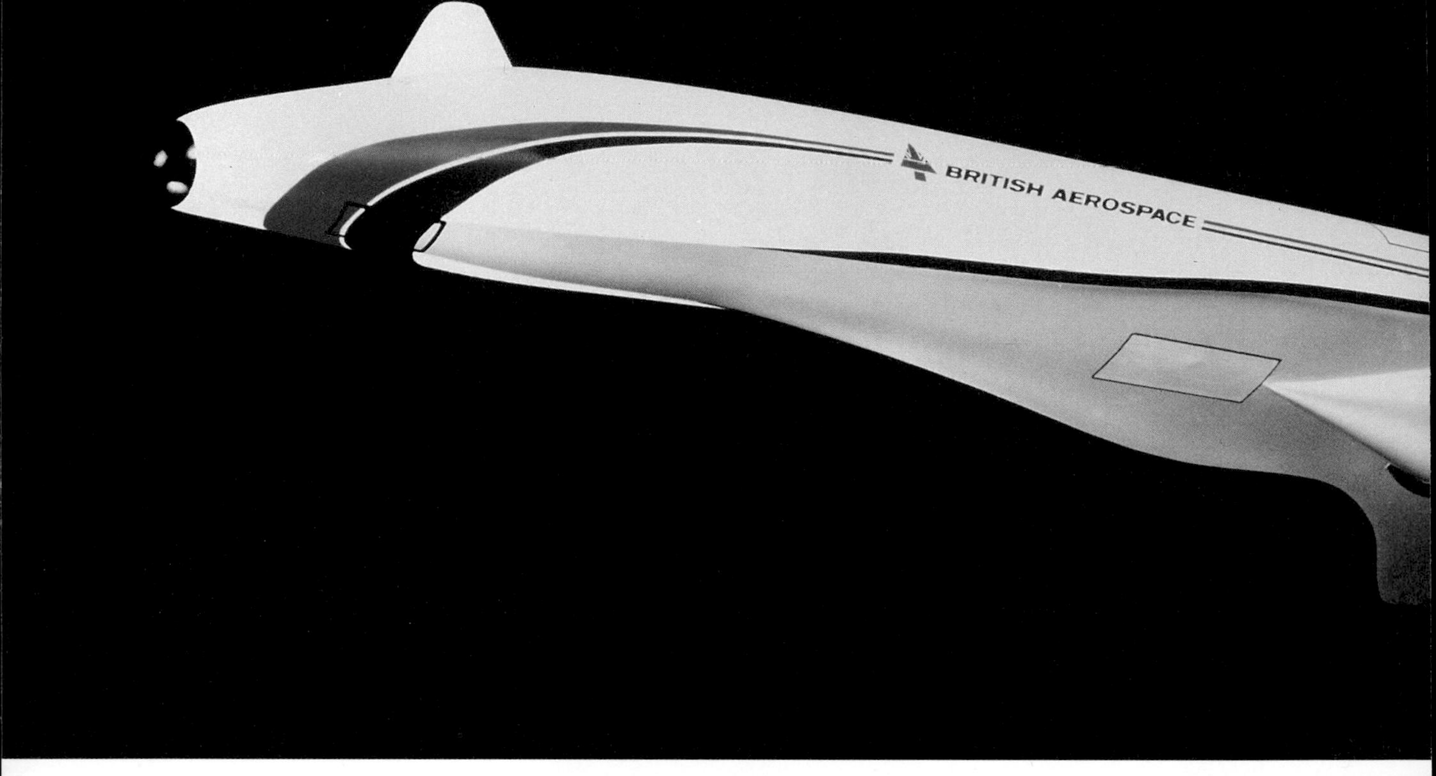

Top: HOTOL (Horizontal Takeoff and Landing) is a spaceplane being designed by British Aerospace.

Bottom: Sometime in the twenty-first century, engineers will develop a true spaceplane *(1)* that can take off and land horizontally like today's airplanes. Several countries are designing such vehicles.

Opposite: NASA is studying these three designs for Single-Stage-to-Orbit reusable launch vehicles. With a so-called lifting body design or a wing body design, the planes would take off and land vertically – just like the rockets in old science-fiction movies!

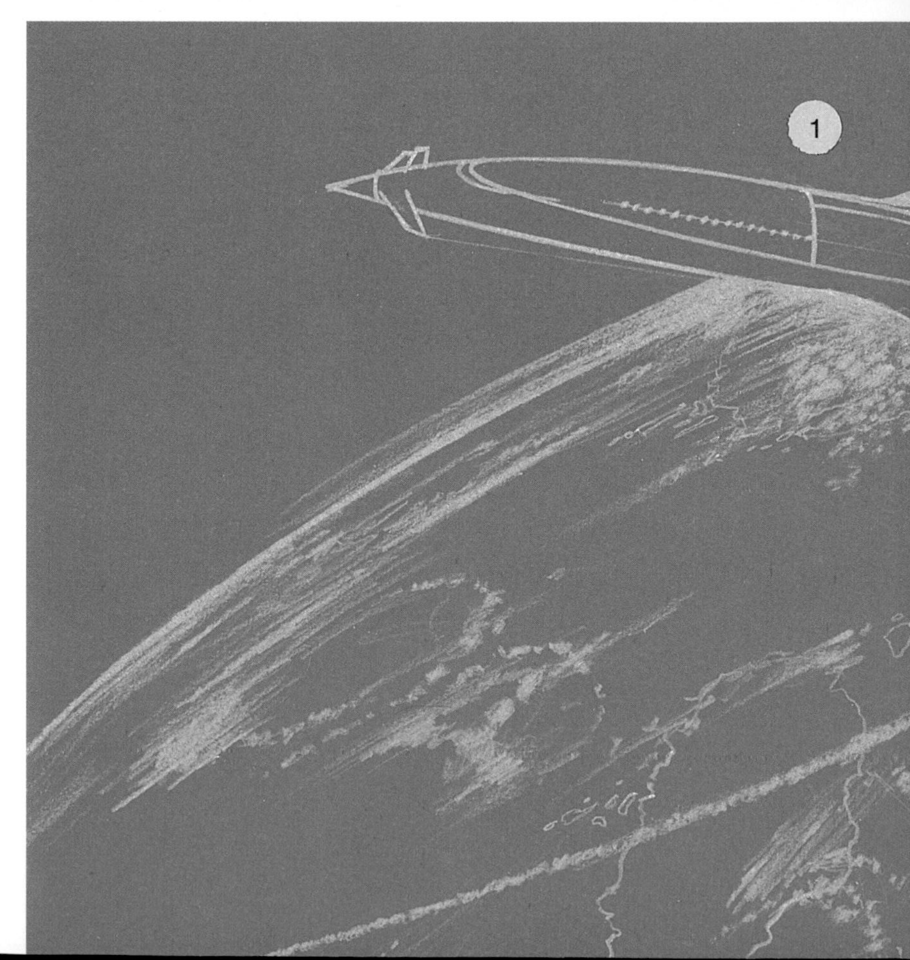

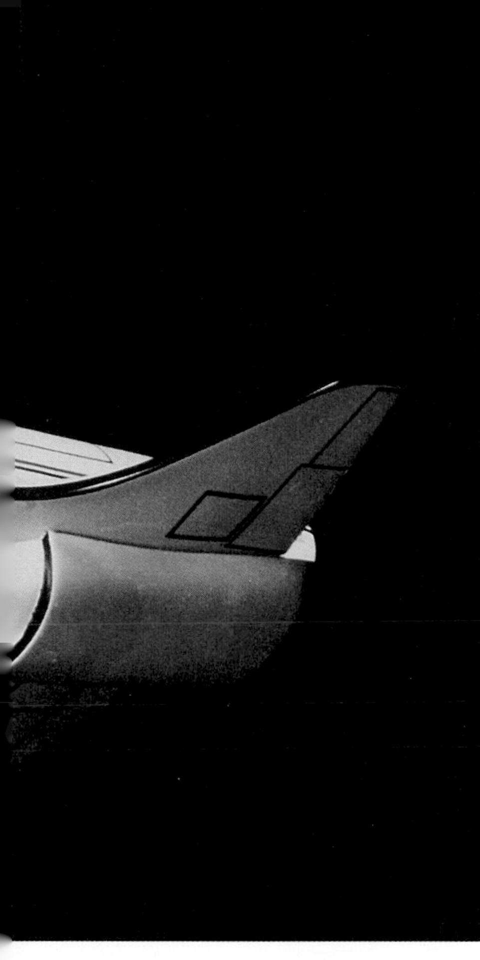

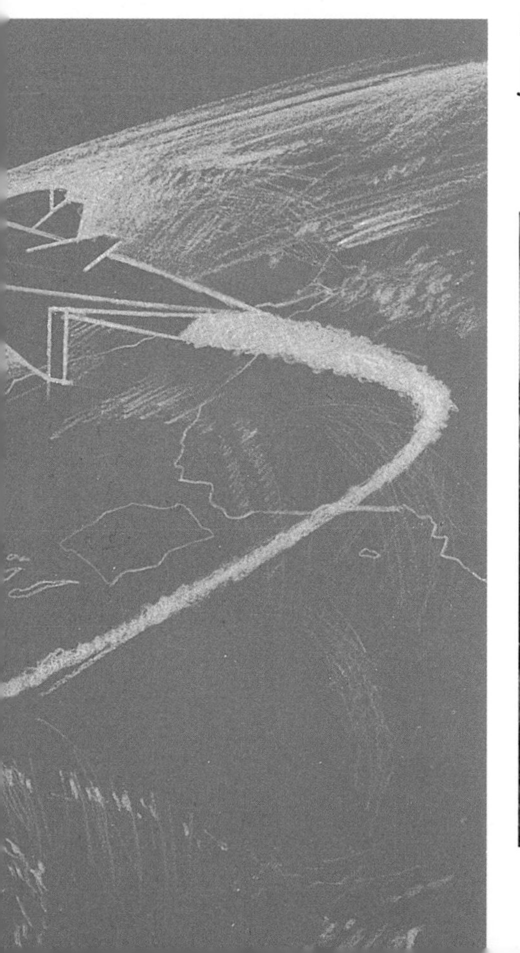

Reusable Launch Vehicles

Although space shuttles are reusable, a large part of the rocket that sends it into space cannot be used again because it burns up in the atmosphere. In addition, the shuttles are based on technology that is now about two decades old. Engineers and designers believe that simpler, completely reusable launch systems can be created.

In the United States, NASA is working with the aerospace industry to explore the idea of SSTO or "Single-Stage-to-Orbit." Instead of using different rocket engines, or stages, that are discarded during launch, an SSTO rocket would blast into orbit on one set of reusable engines. Such designs would be safer and much less expensive to operate.

In time, true spaceplanes will also be developed. Spaceplanes will take off and land at runways just as airplanes do today.

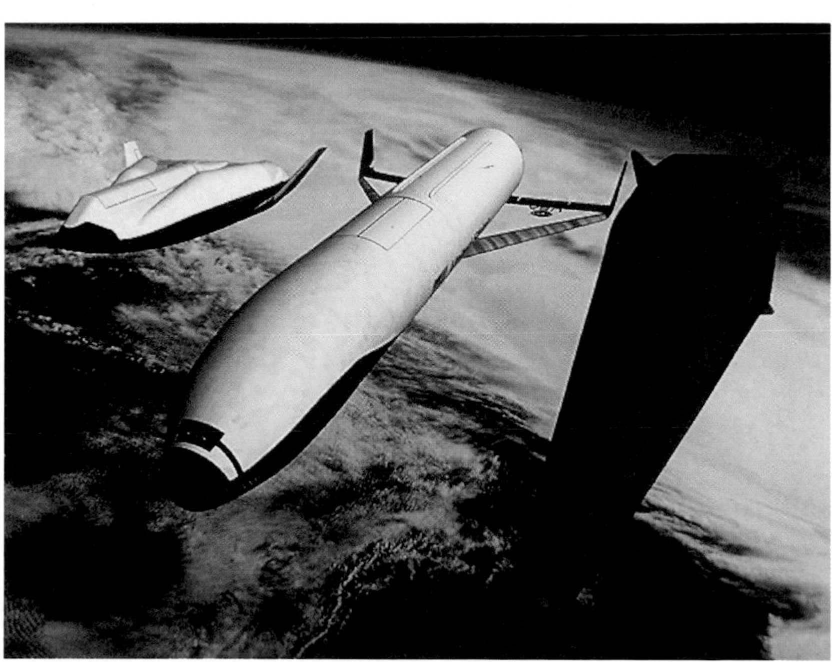

A Cosmic Home

A space station is a permanent satellite in Earth orbit where people can live and work, not just for a day or two, but for months or even years at a time. Advanced space stations could have power stations to gather solar energy, laboratories to carry out experiments, factories to make precision instruments, and observatories to study the cosmos.

Three small space stations have already been placed in orbit – the Russian *Salyut* (1971), the U.S. *Skylab* (1973), and the Russian *Mir* (1986). Astronauts and cosmonauts have lived and worked on the space stations for long periods of time before returning to Earth. Some cosmonauts have remained in space over a year at a time.

The United States, Russia, European nations, and Japan have agreed to work together to build the International Space Station. Russia's *Mir* will form the basis of this station. If all goes according to plan, the International Space Station will be in operation by 2002.

Opposite, top: In this view from the future, a U.S. space shuttle docks with the newly completed International Space Station.

Opposite, bottom, left: In February 1995, the space shuttle *Discovery* inspected the Russian *Mir* space station in a close fly-by. This set the stage for a docking of the two spacecraft later that year.

Opposite, bottom, right: Cosmonaut Valeriy Polyakov, who boarded *Mir* on January 8, 1994, looks out *Mir*'s window during *Discovery*'s fly-by.

? *Gravity – where would we be without it?*

The Moon's gravitational pull is only one-sixth that of Earth's. Similarly, a space station orbiting Earth could be made to spin to produce a gravity-like force, but it would be quite weak compared with Earth's gravity. What effect would a lifetime of weak gravity have on people? What would happen to babies born under such conditions? Would they ever be able to visit Earth with its higher gravity? These are questions scientists cannot yet answer.

Into the Beyond

Someday, space stations could become Earth's launch pads for piloted flights to other planets. A spaceship built at a space station in Earth orbit is already moving very fast. Launching it to the Moon and planets would be easier than an Earth-launch because its engine would need less fuel. Also, a spaceship built in orbit won't be traveling through Earth's atmosphere, so bad weather would never delay a launch. And because there is no air resistance in space, a spaceship built in orbit will not need to have a pointed design.

It is from a space station that humans may depart for Mars and the asteroids. Space stations may also house huge settlements. Each space station would be a self-contained world capable of holding thousands of people, who would make it their home.

Opposite: Hollowing out an asteroid would create a strong shell for an artificial world.

Inset: An advanced space station would have a ring *(1)* that rotates to produce artificial gravity. To anyone in the ring, "down" would be toward the outside of the ring and "up" would be toward the central axle. Fuel and other supplies could be stored in another unit *(2)*. Arriving spacecraft *(3)* could dock to take on supplies or unload cargo.

! *Air – a weighty subject!*

In 1644, Evangelista Torricelli, an Italian scientist, proved that air has weight. From his experiment, scientists began to realize that air exists only near Earth's surface. Beyond Earth was a void. Torricelli, in a roundabout way, had discovered outer space. Anyone going beyond Earth's atmosphere must take their own air supply along.

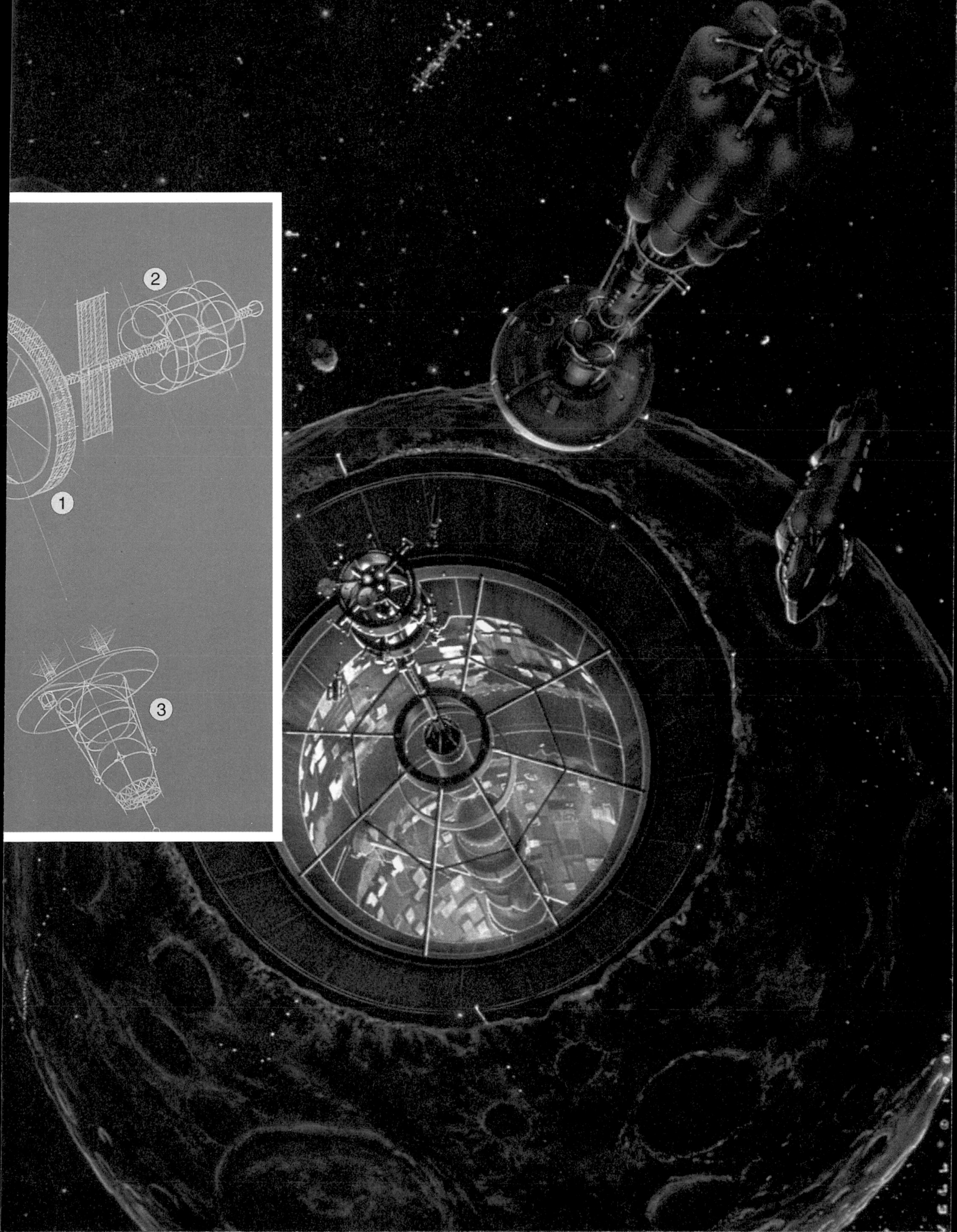

2
1
3

Mapping New Worlds

The world's space programs are bringing us into ever closer touch with the rest of the Solar System. Before humans visited the Moon, probes had already mapped it in detail. Now the same pattern is holding true for Mars and the other planets. Unpiloted space probes – *Mariner*, *Venera*, *Pioneer*, and *Voyager* – have visited the various planets and sent back data.

The probes have mapped Mercury, Venus, and Mars. The four giant planets, Jupiter, Saturn, Uranus, and Neptune, and their satellites have also been studied. They have been visited by the world's most successful probe, *Voyager 2*. Only Pluto has yet to be studied by probes.

Opposite: Early in the twenty-first century, the *Pluto Express* probe will race past the Solar System's most distant planet. Scientists are exploring the possibility of a "Fast Flyby" of Pluto and its moon by 2010.

Below: In the summer of 1997, the Mars *Pathfinder* lander will touch down on the Red Planet. It will then deploy the smallest rover yet sent to another world – it has a mass of just 35 pounds (16 kg)! The rover will send back images and scientific data as it travels the Martian landscape.

! Cosmic rays – no laughing matter!

Cosmic rays bombard Earth from every direction. On Earth, the atmosphere protects us from most of the cosmic rays. In space, cosmic rays are more dangerous. If space settlements house thousands of people, how will they be protected from this radiation? Perhaps the settlements will be buried under the ground and shielded by a thick layer of lunar soil.

The Cosmos Above the Clouds

Telescopes have long been one of the best ways to view the cosmos. But Earth-based telescopes all have the same problem – Earth's atmosphere. While the air surrounding our planet protects us from meteoroids and solar radiation, it can also distort our view of distant objects in space.

A telescope outside the atmosphere is not affected by weather, temperature change, air pollution, or city lights. It can see farther, more clearly, and in greater detail than Earth-based scopes. It can also detect *all* the radiation given off by distant objects, not just the radiation that pierces Earth's atmosphere.

Space telescopes have already detected black holes and warm, dusty clouds around stars that might contain planets. With the launch of the Hubble Space Telescope in 1990, Earth's "eyes" on the stars are open wider than ever.

Opposite: Early in the twenty-first century, a winged Single-Stage-to-Orbit launch vehicle may pay a visit to the Hubble Space Telescope.

The Moon: Earth's Sister-World

Space exploration does not just mean going farther and farther into space. It also means exploring where humans have been before, and doing it more carefully and with greater cooperation among the nations of Earth.

Someday, humans will build mining stations on the Moon. From these stations, minerals will be obtained to build other structures in near-space. An observatory could be built on the far side of the Moon. Underground lunar cities may eventually house millions of people. The Moon will truly be a sister-world of Earth.

Opposite: Large bases on the Moon will be fashioned from the cratered terrain. A giant radio telescope fills a bowl-shaped crater *(1)*. Living quarters *(2)* house scientists near the rim. In the harsh conditions on the lunar surface, occupants of such outposts will need to create their own food, air, and water.

Inset: In this artist's concept of an astronaut exploring the Moon's surface, a reflection of the brightly lit lunar landscape gleams in the astronaut's visor.

! ***Flights of fancy!***

Writers have been imagining flights to the Moon since the time of the Roman Empire. They have told stories of adventurers being carried to the Moon by spirits, being blown there by waterspouts, flying there in chariots dragged by huge geese, and being drawn up by dew. In days gone by, Edgar Allan Poe wrote of space travel in a balloon, Jules Verne shot his explorers out of a giant cannon, and H. G. Wells imagined an anti-gravity device.

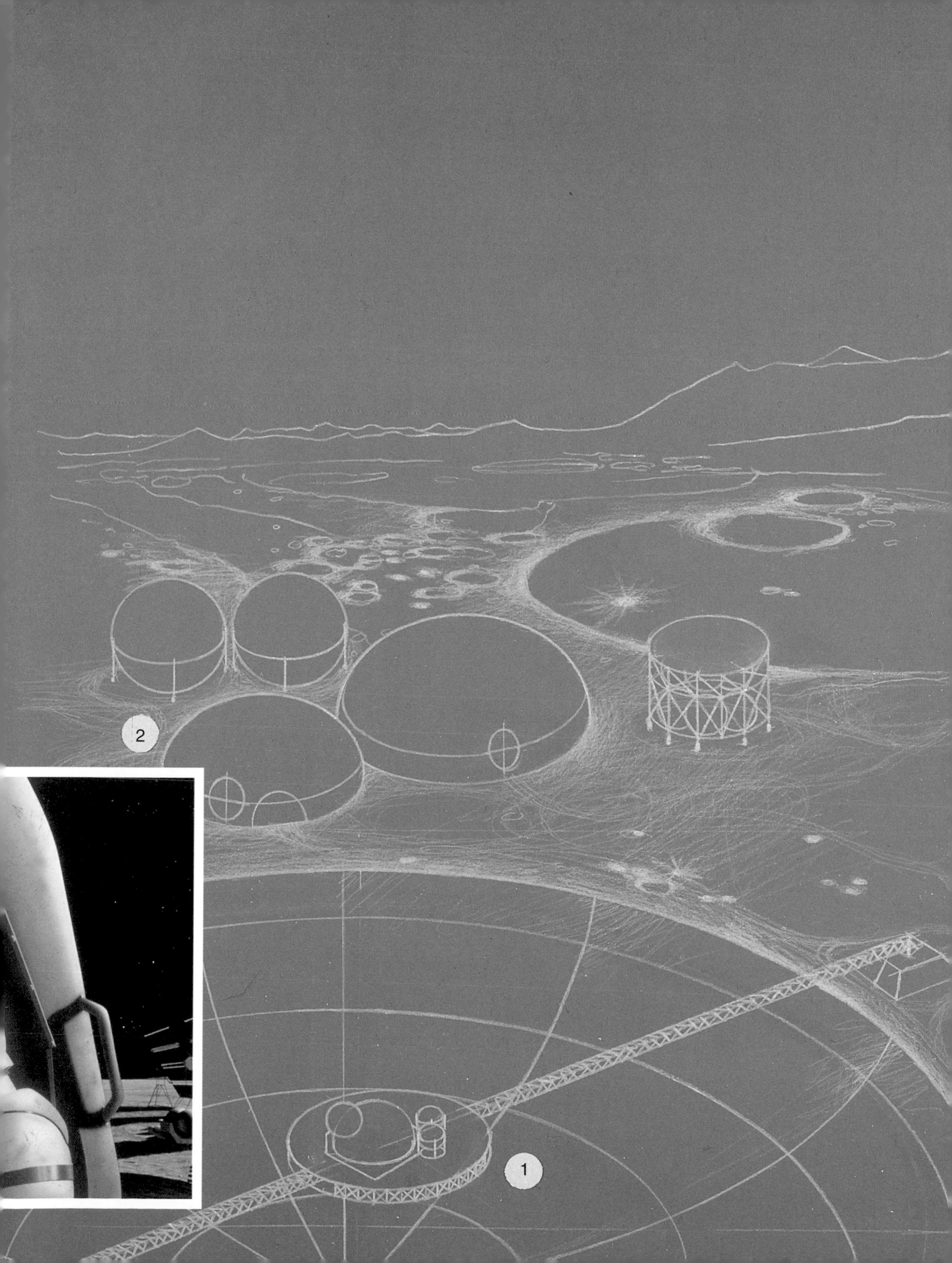
2
1

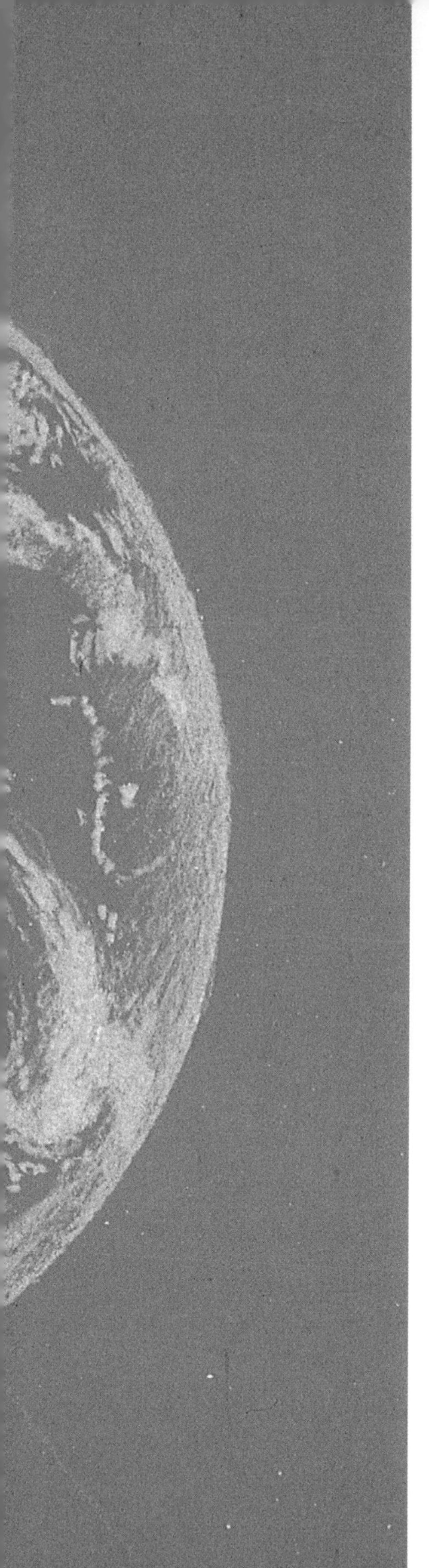

United in Space

One aspect of global space exploration is that humanity is presented with a project so large and important that individual nations cannot carry it through by themselves. The project provides an opportunity for many nations to work together, cooperatively, toward a common goal.

The vast distances of space make Earth seem small. Space contains vast amounts of materials, energy, and scientific data that the nations of Earth could cooperatively pursuc. Because of space exploration, people may eventually come to identify themselves as fellow Earthlings, or, better yet, humans, and nothing else. Maybe, at last, people of all nations can learn to share this home of ours.

Left: All Earthly beings are actually riding on a huge spaceship right now. Traveling at 65,865 miles (106,000 kilometers) per hour as it circles the Sun, the planet Earth is a delicate balance of atmosphere, geology, and living creatures.

Fact File: Stairway to Space

In 1967, the United Nations passed the Outer Space Treaty. This treaty states that the Moon and all the rest of space are open to exploration by all nations and belong to none. But in the 1950s and for most of the 1960s, space exploration was primarily a race – a "space race" – between the world's chief Space Age rivals, the former Soviet Union and the United States. But this sense of competition has faded over time. Today, many nations work together on space projects.

One of the best examples of nations working together to explore space is the European Space Agency, or ESA. Founded in 1975 by a group of European nations, with Canada participating as an associate member, ESA pools the resources and knowledge of its member nations.

Today, all the major space-exploring nations launch satellites for countries without rockets of their own. Both Russia and the United States have played host to astronauts or cosmonauts from many different countries. More and more, nations are working together to make the exploration of space truly a global endeavor. Following is a listing of steps on the "stairway to space," showing how nations first entered global space programs.

Major "Firsts" for Nations Participating in Global Space Programs

GERMANY
October 3, 1942
V-2 rocket (first true liquid-fueled rocket) used toward end of World War II.

RUSSIA (former Soviet Union)
October 4, 1957
Sputnik 1 (first artificial satellite) launched. On April 12, 1961, Yuri Gagarin became the first person in space.

UNITED STATES
January 31, 1958
Explorer 1 (independent launch of artificial satellite). On May 5, 1961, aboard *Gemini 3*, Alan Shepard became the first U.S. astronaut in space. On July 20, 1969, Neil Armstrong of *Apollo 11* set foot on the Moon. On February 6, 1995, space shuttle *Discovery* passed within 36 feet (11 m) of Russia's *Mir* space station.

UNITED KINGDOM
April 26, 1962
Ariel 1 (artificial satellite launched by United States). On October 28, 1971, from the Australian pad at Woomera, the U.K. launched its first artificial satellite independently.

CANADA
September 29, 1962
Alouette 1 (artificial satellite launched by United States). On October 5, 1984, aboard the U.S. space shuttle *Challenger*, Marc Garneau became the first Canadian in space.

ITALY
December 15, 1964
San Marco 1 (artificial satellite launched by United States).

FRANCE
November 26, 1965
A-1 (Asterix) (independent launch of artificial satellite). French *spationautes* rode into space aboard the Soviet *Salyut* space station in 1982 and aboard a U.S. space shuttle in 1985.

AUSTRALIA
November 29, 1967
Wresat (artificial satellite launched by United States).

WEST GERMANY
November 8, 1969
Azur (artificial satellite launched by United States). In 1983 and 1985, three West German astronauts flew aboard U.S. space shuttles.

JAPAN
February 11, 1970
Osumi (independent launch of artificial satellite).

CHINA
April 24, 1970
China 1 (Tung-Fang-Hung) (independent launch of artificial satellite).

THE NETHERLANDS
August 30, 1974
ANS 1 (Astronomical Netherlands Satellite) (artificial satellite launched by United States). On October 30, 1985, aboard U.S. shuttle *Challenger*, Wubbo Ockels became the first Dutch astronaut in space.

INDIA
April 19, 1975
Aryabhata (artificial satellite launched by the former Soviet Union). On July 18, 1980, India made its first independent artificial satellite launch.

INDONESIA
July 8, 1976
Palapa 1 (artificial satellite launched by United States).

CZECHOSLOVAKIA
March 2, 1978
Vladimir Remek was a guest cosmonaut aboard Soviet *Salyut* space station.

BULGARIA
April 10, 1978
Georgi Ivan Ivanov was a guest cosmonaut aboard Soviet *Soyuz*.

POLAND
June 27, 1978
Miroslaw Hermaszewski was a guest cosmonaut aboard Soviet *Salyut* space station.

EAST GERMANY
August 26, 1978
Sigmund Jahn was a guest cosmonaut aboard Soviet *Salyut* space station.

HUNGARY
May 26, 1980
Bertalan Farkas was a guest cosmonaut aboard Soviet *Salyut* space station.

VIETNAM
July 23, 1980
Pham Tuan was a guest cosmonaut aboard Soviet *Salyut* space station.

CUBA
September 18, 1980
Arnaldo Tamayo Mendez was a guest cosmonaut aboard Soviet *Salyut* space station.

MONGOLIA
March 22, 1981
Jugderdemidyin Gurragcha was a guest cosmonaut aboard Soviet *Salyut* space station.

ROMANIA
May 15, 1981
Dumitru Prunariu was a guest cosmonaut aboard Soviet *Salyut* space station.

SPAIN
November 15, 1984
Intasat (artificial satellite launched by United States).

SAUDI ARABIA
February 8, 1985
Arabsat 1 (artificial satellite launched by ESA). On June 18, 1985, *Arabsat 2* was launched into orbit from U.S. shuttle *Discovery* by guest astronaut Prince Sultan Abdul Aziz Al-Saud.

MEXICO
June 18, 1985
Morelos 1 (artificial satellite launched from U.S. space shuttle *Discovery*). On November 27, 1985, aboard the U.S. shuttle *Atlantis*, Rodolfo Neri Vela became the first Mexican astronaut in space.

BRAZIL
February 8, 1985
Brasilsat 1 (artificial satellite launched by ESA).

SWEDEN
February 22, 1986
Viking (artificial satellite launched by ESA).

SYRIA
July 22, 1987
Mohammed Faris was a guest cosmonaut aboard Soviet *Mir* space station.

AFGHANISTAN
August 29, 1988
M. D-G. Masum was a guest cosmonaut aboard Soviet *Soyuz*.

More Books about Global Space Programs

Album of Space Flight. McGowen (Macmillan)
Exploring Outer Space: Rockets, Probes, and Satellites. Asimov (Gareth Stevens)
From Carriage to Spaceship. Mikhalkov (Import/Progress)
Modern Astronomy. Asimov (Gareth Stevens)
Space Explorers. Asimov (Gareth Stevens)
Sputnik to Space Shuttle: The Complete Story of Space Flight. Nicholson (Dodd, Mead)
The 21st Century in Space. Asimov (Garcth Stevens)

Video

Astronomy Today. (Gareth Stevens)

Places to Visit

You can explore the Universe without leaving Earth. Here are some museums and centers where you can find many different kinds of space exhibits.

NASA Lyndon B. Johnson Space Center
2101 NASA Road One
Houston, TX 77058

International Women's Air and Space Museum
1 Chamber Plaza
Dayton, OH 45402

The Space and Rocket Center
and Space Camp
One Tranquility Base
Huntsville, AL 35807

San Diego Aero-Space Museum
2001 Pan American Plaza – Balboa Park
San Diego, CA 92101

Australian Museum
6-8 College Street
Sydney, NSW 2000 Australia

Astrocentre
Royal Ontario Museum
100 Queen's Park
Toronto, Ontario M5S 2C6

Places to Write

Here are some places you can write for more information about space exploration. Include your full name and address so the organization can write back to you.

For a free copy of *Space for Women*:
Harvard-Smithsonian Center
for Astrophysics
Publications Department, MS-28, Dept. P
60 Garden Street
Cambridge, MA 02138

Canadian Space Agency
Communications Department
6767 Route de L'Aeroport
Saint Hubert, Quebec J3Y 8Y9

To subscribe to a children's astronomy magazine:
Odyssey
Cobblestone Publishing, Inc. Dept. P
7 School Street
Peterborough, NH 03458-1454

Sydney Observatory
P. O. Box K346
Haymarket 2000
Australia

Glossary

astronauts: men and women from many countries who travel beyond the atmosphere of Earth.

atmosphere: the layer of gases surrounding a planet, star, or moon. Most satellites orbit just outside Earth's atmosphere to avoid burning up from friction produced when there is contact with the atmosphere.

billion: the number represented by 1 followed by nine zeroes – 1,000,000,000. In some countries, this number is called "a thousand million." In these countries, one billion would then be represented by 1 followed by twelve zeroes – 1,000,000,000,000 – a million million.

black hole: an object in space caused by the explosion and collapse of a star. The object is so tightly packed that not even light can escape the force of its gravity.

cosmonauts: men and women from Russia or the former Soviet Union who travel beyond the atmosphere of Earth.

erosion: the wearing away of a substance by such forces as wind and water.

European Space Agency (ESA): an organization founded in 1975, pooling the resources of several European countries and Canada for joint research and exploration of space.

gravity: the force that causes objects like planets and their moons to be drawn to one another.

NASA: the space agency in the United States – the National Aeronautics and Space Administration.

navigational: having to do with planning or directing the course or path of a craft.

observatory: a building or other structure designed for watching and recording celestial objects and events.

rocket: a missile used to launch satellites, probes, shuttles, and other craft into space.

rocketry: the study of, experimentation with, or use of rockets.

satellite: a smaller body that orbits a larger body. *Sputnik 1* and *2* were Earth's first artificial satellites. The Moon is Earth's natural satellite.

space shuttle: a reusable spacecraft launched into space by a rocket but capable of returning to Earth under its own power. The first space shuttle, *Columbia*, was launched in 1981 by the United States.

SSTO: Single-Stage-to-Orbit rocketry, an idea that instead of using different rocket engines, or stages, that are thrown away during launch, space explorers can use one set of reusable engines.

Index

Born in 1920, Isaac Asimov came to the United States as a young boy from his native Russia. As a young man, he was a student of biochemistry. In time, he became one of the most productive writers the world has ever known. His books cover a spectrum of topics, including science, history, language theory, fantasy, and science fiction. His brilliant imagination gained him the respect and admiration of adults and children alike. Sadly, Isaac Asimov died shortly after the publication of the first edition of *Isaac Asimov's Library of the Universe*.

The publishers wish to thank the following for permission to reproduce copyright material: front cover, NASA; 4-5, Matthew Groshek/© Gareth Stevens, Inc.; 5, NASA; 6-7, Matthew Groshek/© Gareth Stevens, Inc.; 7, NASA; 8-9, Matthew Groshek/© Gareth Stevens, Inc.; 9, Courtesy of Marconi Space Systems; 10, NASA; 10-11 (upper), Courtesy of COMSAT; 10-11 (lower), Matthew Groshek/© Gareth Stevens, Inc.; 13, NASA; 14-15 (upper), Courtesy of British Aerospace; 14-15 (lower), Matthew Groshek/© Gareth Stevens, Inc.; 15, 17 (all), NASA; 18-19, Matthew Groshek/© Gareth Stevens, Inc.; 19, © Mark Maxwell 1989; 20, 21, Jet Propulsion Laboratory; 23, NASA; 24-25, Mike Stovall and Mark Dowman/NASA; 25, 26-27, Matthew Groshek/© Gareth Stevens, Inc.